AF575717

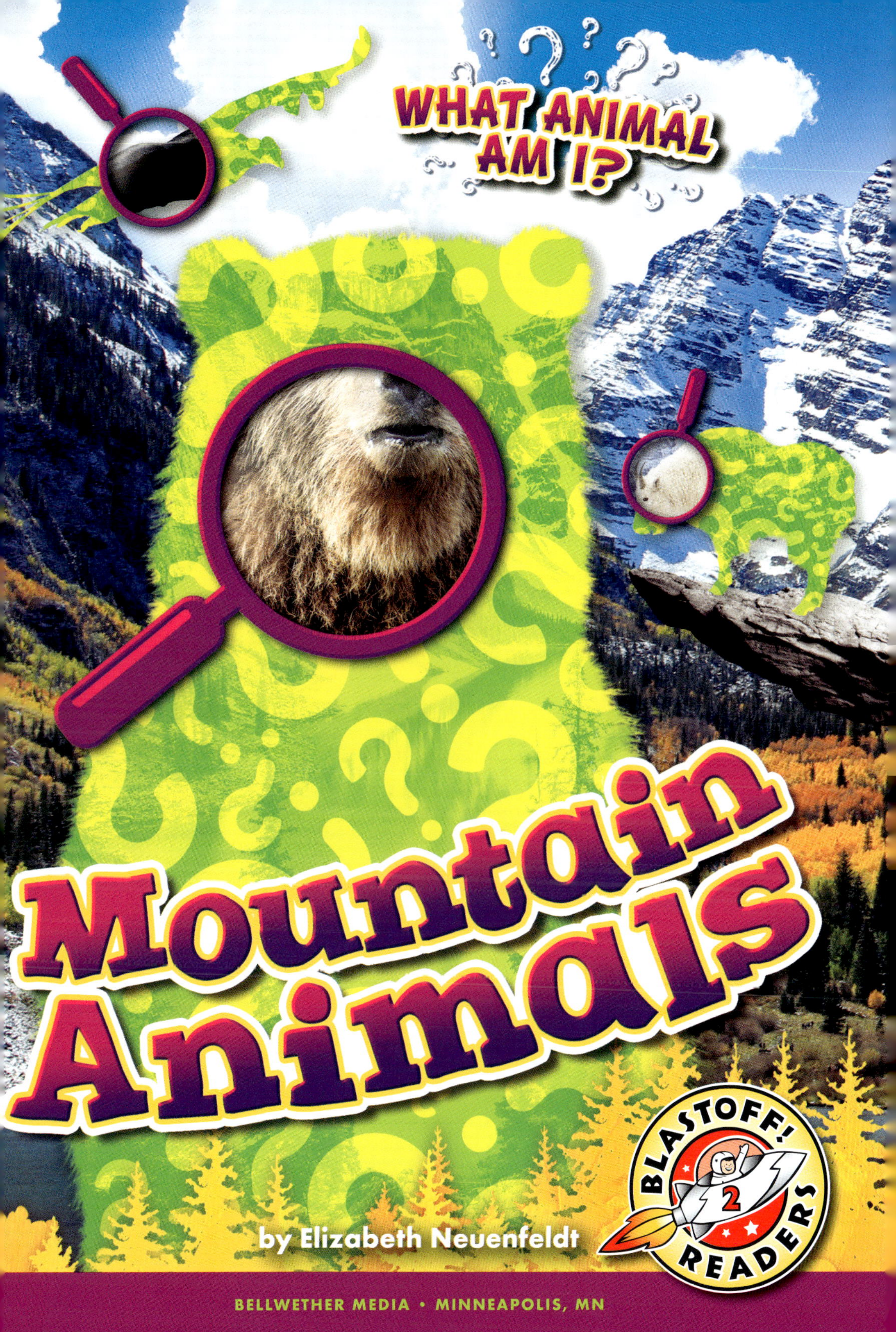
WHAT ANIMAL AM I?
Mountain Animals
by Elizabeth Neuenfeldt
BLASTOFF! READERS
2
BELLWETHER MEDIA • MINNEAPOLIS, MN

Blastoff! Readers are carefully developed by literacy experts to build reading stamina and move students toward fluency by combining standards-based content with developmentally appropriate text.

Level 1 provides the most support through repetition of high-frequency words, light text, predictable sentence patterns, and strong visual support.

Level 2 offers early readers a bit more challenge through varied sentences, increased text load, and text-supportive special features.

Level 3 advances early-fluent readers toward fluency through increased text load, less reliance on photos, advancing concepts, longer sentences, and more complex special features.

★ **Blastoff! Universe**

Reading Level

Grade K

Grades 1–3

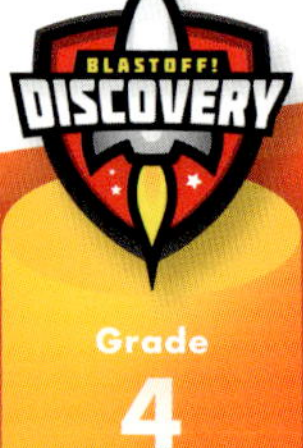

Grade 4

This edition first published in 2023 by Bellwether Media, Inc.

Library of Congress Cataloging-in-Publication Data

Names: Neuenfeldt, Elizabeth, author.
Title: Mountain animals / by Elizabeth Neuenfeldt.
Description: Minneapolis, MN : Bellwether Media, [2023] | Series: What animal am I? | Includes bibliographical references and index. | Audience: Ages 5-8 | Audience: Grades 2-3 | Summary: "Relevant images match informative text in this introduction to different mountain animals. Intended for students in kindergarten through third grade" -- Provided by publisher.
Identifiers: LCCN 2022009376 (print) | LCCN 2022009377 (ebook) | ISBN 9781644877296 (library binding) | ISBN 9781648347757 (ebook)
Subjects: LCSH: Mountain animals--Juvenile literature.
Classification: LCC QL113 .N48 2023 (print) | LCC QL113 (ebook) | DDC 591.7--dc23/eng/20220303
LC record available at https://lccn.loc.gov/2022009376
LC ebook record available at https://lccn.loc.gov/2022009377

Editor: Rachael Barnes Designer: Brittany McIntosh

Printed in the United States of America, North Mankato, MN.

Table of Contents

Welcome to the Mountains!

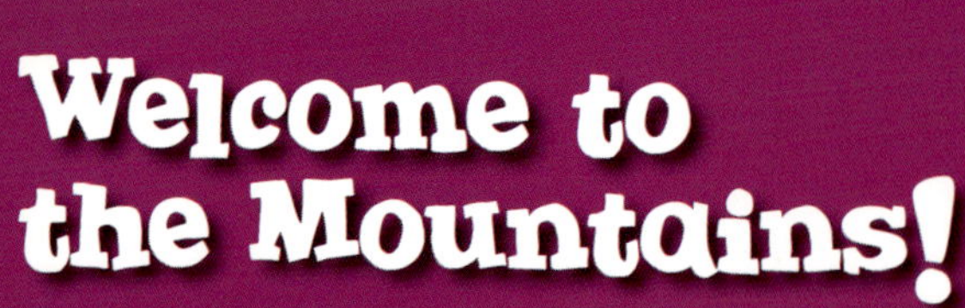

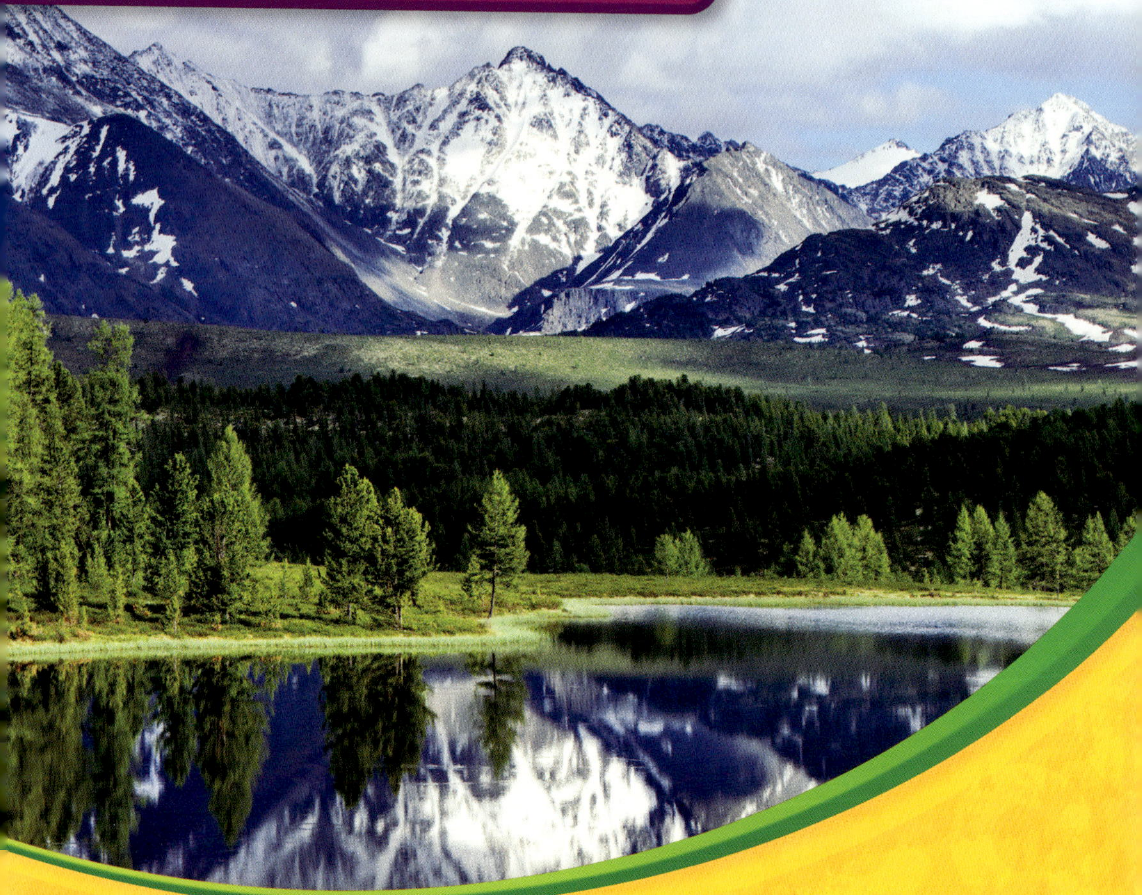

Mountains are giant, rocky **landforms**. They can be very cold and snowy.

Many animals live well in this tough **biome**!

A Spotted Cat

I am a large **mammal** with thick, spotted fur. I have wide, furry paws!

I wind my long tail around my body to stay warm. What animal am I?

I am a snow leopard!
I am a **carnivore**.

I leap across snowy cliffs to follow **prey**. My powerful legs help me **pounce**!

Snow Leopard Food
sheep
deer
hares

One Big Bird

I am one of the largest flying birds! My huge wings help me fly over mountains.

Dark feathers cover my body, but my head is **bald**. What animal am I?

Andean Condor Food
carrion
small birds
rabbits

I am an Andean condor! I eat **carrion** and small living animals.

I have excellent eyesight. While I fly, I easily spot food below me!

A Cliff Climber

I am a large mammal with horns. My thick, white **coat** keeps me warm.

My coat also blends in among snow-covered rocks! What animal am I?

I am a mountain goat! I eat **lichens** and grass on steep cliffs.

My padded **hooves** help me climb!

Mountain Goat Food
lichens
evergreen trees
grass

A Sleepy Mammal

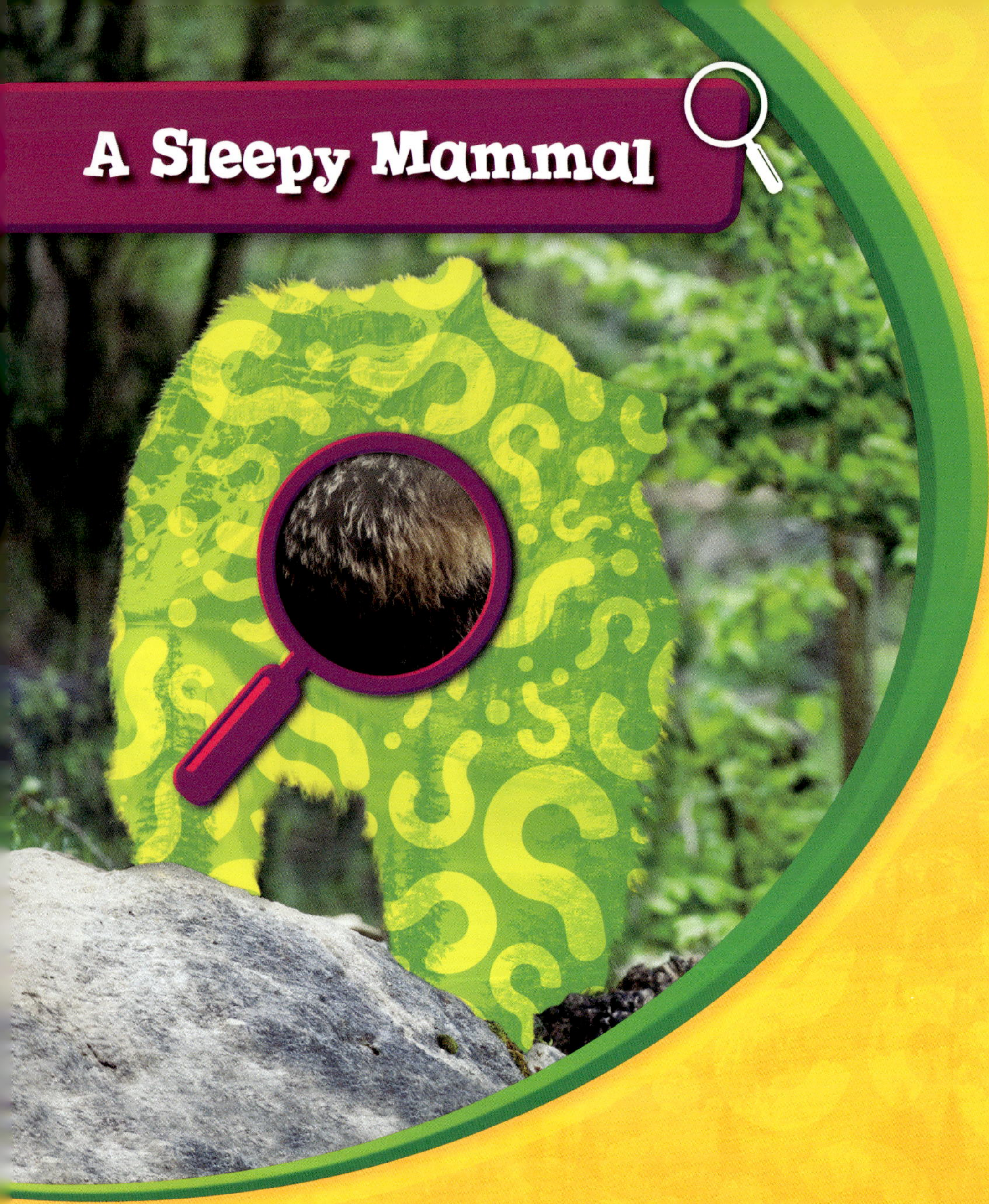

I am a giant mammal with large paws and claws.

When winter comes, I **hibernate** inside my **den**. What animal am I?

Brown Bear Food
fish
berries
nuts

I am a brown bear! I eat a lot of food in summer and fall. This helps me hibernate through winter.

The mountains have many tough animals!

hibernating in a den

Glossary

bald—having no hair or feathers

biome—a large area with certain plants, animals, and weather

carnivore—an animal that only eats meat

carrion—the rotting meat of a dead animal

coat—the hair or fur covering some animals

den—a sheltered place

hibernate—to spend the winter sleeping or resting

hooves—hard coverings on the feet of some animals

landforms—natural features on Earth's surface; mountains and valleys are landforms.

lichens—plantlike living things that grow on rocks

mammal—a warm-blooded animal that has a backbone and feeds its young milk

pounce—to suddenly jump on something to catch it

prey—animals that are hunted by other animals for food

To Learn More

AT THE LIBRARY

Hicks, Dwayne. *That's a Mountain!* New York, N.Y.: Gareth Stevens Publishing, 2022.

Rissman, Rebecca. *Kings of the Mountains.* North Mankato, Minn.: Capstone, 2018.

Sabelko, Rebecca. *Mountains.* Minneapolis, Minn.: Bellwether Media, 2022.

ON THE WEB

FACTSURFER

Factsurfer.com gives you a safe, fun way to find more information.

1. Go to www.factsurfer.com.

2. Enter "mountain animals" into the search box and click 🔍.

3. Select your book cover to see a list of related content.

Index

The images in this book are reproduced through the courtesy of: Volodymyr Burdiak, front cover (bear), p. 21; aaltair, front cover (condor); BearFotos, front cover (goat), p. 11 (left); liyang0518, front cover (background); Marti Bug Catcher, pp. 3, 23; Valerii_M, p. 4; Fedor Selivanov, p. 5; Dennis W Donohue, p. 6; Ondrej Prosicky, pp. 7 (left), 9 (top right); clarst5, p. 7 (right); slowmotiongli, p. 8; Jeff Wendorff/ Getty Images, pp. 8-9; JIANG TIANMU, p. 9 (top left); Fabio Nodari, p. 9 (top middle); All Canada Photos/ Alamy, p. 10; Vladimir Wrangel, p. 11 (right); Lapis2380, p. 12 (top left); David Havel, p. 12 (top middle and top right); Fominayaphoto, pp. 12-13; bukentagen, p. 13; Images by Dr. Alan Lipkin, p. 14; Danita Delimont, p. 15 (left); Scalia Media, p. 15 (right); franzfoto.com/ Alamy, p. 16; BGSmith, pp. 16-17; andyKRAKOVSKI/ iStockphoto, p. 17 (top left); Serp, p. 17 (top middle); Heinz Staudacher, p. 17 (top right); Peter Wey, p. 18; angellodeco, p. 19 (left); Azahara Perez, p. 19 (right); Beat J Korner, p. 20 (top left); Zabavna, p. 20 (top middle); Jiri Hera, p. 20 (top right); Henk Bogaard, pp. 20-21; Martin Bergsma, p. 22.